WEIRD PLANTS

WEIRD
PLANTS

CHRIS THOROGOOD

Kew Publishing
Royal Botanic Gardens, Kew

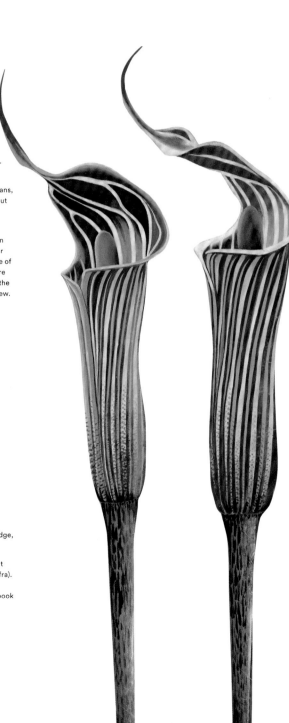

Royal
Botanic Kew
Gardens

© The Board of Trustees of the Royal Botanic Gardens, Kew, 2018
Text and images © Chris Thorogood unless stated otherwise

Reprinted 2018

First published in 2018 by the Royal Botanic Gardens, Kew,
Richmond, Surrey, TW9 3AB, UK
www.kew.org

ISBN 978 1 84246 662 9

Distributed on behalf of the Royal Botanic Gardens, Kew
in North America by the University of Chicago Press,
1427 East 60th St, Chicago, IL 60637, USA

British Library Cataloguing in Publication Data
A catalogue record for this book is available from the British Library.

Design and page layout: Ocky Murray
Copy-editing: Christina Harrison
Proofreading: Matthew Seal
Production Management: Georgie Hills

For information or to purchase all Kew titles please visit
shop.kew.org/kewbooksonline or email publishing@kew.org

Kew's mission is to be the global resource in plant and fungal knowledge,
and the world's leading botanic garden.

Kew receives approximately one third of its funding from Government
through the Department for Environment, Food and Rural Affairs (Defra).
All other funding needed to support Kew's vital work comes from
members, foundations, donors and commercial activities, including book
sales.

Printed and bound in Great Britain by Gomer Press

MIX
Paper from
responsible sources
FSC® C114687

CONTENTS

AUTHOR'S PREFACE

....

I have been fascinated by plants for as long as I can remember. I spent much of my childhood growing bizarre and unusual plants and illustrating them. I have carried my passion for plants and painting with me ever since, and I've been fortunate enough to see some rare and extraordinary plants in their natural environments. Nothing quite compares with standing amid a grove of gigantic pitcher plants on a misty mountainside in Borneo, peering into the world's largest flower in the heart of the Malaysian rainforest, or hunting for one of the weirdest of all plants, *Hydnora*, in the deserts of South Africa. Experiences such as these were the inspiration for my paintings, in which I have sought to capture the plants in all their natural splendour in oils.

Some of the weird plants I have illustrated will be very familiar. Most of us remember being fascinated by the formidable jaws of a Venus flytrap as children. But I have also tried to include plants that, surprisingly, we still know very little about: plants completely new to science; parasitic plants, many of which are unknown in cultivation; and even seemingly familiar tropical pitcher plants, which we now know have astoundingly intricate relationships with animals, the likes of which scientists could never have imagined.

Today, fascinating plants and their natural histories still await discovery. Since the preparation of this book, at least one pitcher plant appears to have become extinct in the Philippines through habitat destruction and is now sadly lost to science forever. And this is ultimately at the heart of why I have captured and compiled this unique assortment of botanical oddities: I want to inspire and intrigue you with the plant world, which has always captivated me. I hope that a new generation of botanists and horticulturalists will continue to explore, appreciate and conserve the extraordinary plants with which we share our planet for many future generations.

INTRODUCTION

....

Our green planet is home to a dazzling diversity of flowering plant species, hundreds of thousands of them. From arctic tundra to barren deserts, plants have evolved myriad different strategies to survive under extreme, sometimes hostile conditions; strategies to jostle with the elements, to compete with one another for pollinators and to disperse their seeds.

This book explores the bizarre, the sinister and the mysterious side of the plant world. A world in which plants trick, dupe, steal and even kill: carnivorous plants that drug, drown and consume unsuspecting insect prey; giant pitcher plants that evolved toilets for tree shrews; flowers that mimic rotting flesh to attract pollinating flies, and orchids that duplicitously look, feel and even smell like a female insect to bamboozle sex-crazed male bees. Whilst some of these plants are well-known to science, others are still poorly understood – for example *Rafflesia*, the world's largest flower, is still virtually unknown in cultivation and represents a giant botanical enigma.

In an age of instant information and greater accessibility to plants in the wild than ever before, this book seeks to build context and connection among some of the most extraordinary plants – from the well- known to the enigmatic. It is designed to showcase plant behaviour: the inter-relationships among plants, the inter-dependencies between plants and animals, and the intrigue of plant evolution. Ultimately, it is about the science of weird plants.

Left: The giant, cabbage-like bud of *Rafflesia arnoldii*, the largest flower in the world.

WEIRD PLANTS AROUND THE WORLD

This map indicates approximately where the weird plants featured in this book occur around the globe. More precise distributions for each species are provided in the text.

USA & CENTRAL AMERICA

Trumpet pitcher

Venus flytrap

Mitrastemon

Aristolochia arborea

Parrot pitcher plant

MEDITERRANEAN

Mediterranean arum

Dead horse arum

Maltese fungus

Cistanche

Cytinus

Bee orchid

Tongue orchid

SOUTH AMERICA

Bucket orchid

Marsh pitcher plant

Monkey-face orchid

SOUTHERN AFRICA

Hydnora

Toad cactus

Starfish flower

Welwitschia

Half-mens

Candelabra lily

Bird of paradise

WIDESPREAD

Dutchman's pipe

Snake gourd

Mistletoe

Sundew

Dodder

Broomrape

Devil's guts

BORNEO

Low's pitcher plant

King pitcher plant

Rothschild's slipper orchid

Bat pitcher plant

Fanged pitcher plant

SOUTHEAST ASIA

Titan arum

Rafflesia

Rhizanthes

Durian

Ant plant

INDIAN OCEAN

Traveller's palm

Coco de mer

AUSTRALASIA

Flying duck orchid

Balanophora

New Zealand flax

Sturt's desert pea

SECTION 1

VAMPIRES

Parasitic plants have aroused curiosity for centuries yet they remain one of the most poorly understood groups of all the flowering plants. Indeed much of their evolutionary biology and life history remains a mystery. There are over 4,000 species of parasitic plant and they occur in all major ecosystems from tropical rainforests to arctic tundra. They attach to the roots or stems of other plants – their so-called hosts – from which they extract water and nutrients. Many derive all their nutrition from their hosts and so have lost the features typical of most plants, for example green leaves, stems and even roots. Because botanists previously relied on such features to classify plants, the evolutionary relationships among these parasites have long remained unclear. However, DNA sequencing technology has unearthed some surprises among their genetic relatedness. For example tropical *Rafflesia*, famously the largest flower on Earth, is now established to be closely related to euphorbias, a family of plants with typically minute and inconspicuous flowers.

Left: Parasitic *Rafflesia arnoldii* is the largest flower on Earth, measuring up to an incredible 1.5 m (5 ft) across. The existence of this gigantic 'vampire plant' in the rainforests of Sumatra is probably dependent on tigers, which share its habitat and are the focus of conservation efforts.

13

RAFFLESIA
World's largest flower
Rafflesia spp.

Rafflesia is the largest single flower on Earth. There are about 30 species of *Rafflesia* in the forests of Southeast Asia; the largest is R. *arnoldii* the flowers of which can span up to 1.5 m (5 ft) across and weigh over 10 kg (22 lb). *Rafflesia* completely lacks leaves, roots and stems and spends most of its life embedded within the tissues of its host plant, a tropical vine (*Tetrastigma*) in the grape family. *Rafflesia* has evolved a resemblance to rotting flesh and produces a strong, unpleasant smell to attract pollinating flies from far and wide. Despite its fame for being the world's largest flower, relatively little is known about the plant's life history, and it remains virtually unknown in cultivation. DNA sequencing recently revealed the plant's closest photosynthetic relatives to be the euphorbias which, surprisingly, produce very small and inconspicuous flowers.

Right: *Rafflesia pricei* is one of the best-conserved species of *Rafflesia*. It grows in sub-montane forests that fall within protected areas of North Borneo. It even has a dedicated '*Rafflesia* sanctuary', and is an emblem of eco-tourism and conservation in the area.

Above: *Rafflesia* produces giant, cabbage-like buds, such as the specimen of R. *keithii*, photographed in the rainforests of Sabah, North Borneo.

Following spread: The enormous flowers of R. *keithii*.

RHIZANTHES

Sea monsters in the jungle

Rhizanthes spp.

....

This bizarre plant would look more at home at the bottom of the sea than the rainforest floor. Fungus-like *Rhizanthes* measures about 15 to 40 cm (6 to 16 in) across. Like its relative *Rafflesia,* this rarely seen parasitic plant is devoid of all green pigment, leaves, stems and roots, and is completely dependent upon the roots of tropical vines for its nutrition. There are only four species known to science, all of them occurring in the rainforests of Southeast Asia. Just like *Rafflesia,* the flower resembles rotting meat to attract flies, and even produces heat to disperse its fragrance far and wide across the rainforest. Very little is known about the life history of this curious plant, and like many of the tropical parasitic plants, it is virtually unknown in cultivation.

Right: *Rhizanthes lowii*, native to the rainforests of Borneo.

MITRASTEMON

Tree sap pilferer

Mitrastemon matudae

....

This alien-looking plant is *Mitrastemon matudae*, a poorly known parasite from the mountain forests of Mexico. It only has one close relative, *M. yamamotoi,* which occurs in tropical and subtropical Asia, and both species are relatively little studied. *Mitrastemon matudae* was discovered less than a century ago, and was described to resemble a piece of 'wax art work', with its shiny, milky surface. *Mitrastemon* feeds from the roots of trees in the oak family (Fagaceae). Like *Rafflesia*, this root parasite grows within the tissues of its host plant, only emerging when it blooms. In fact because of this peculiar shared life history, it was once believed to be closely related to *Rafflesia*. Using DNA sequence data, scientists have now established that *Mitrastemon* is more closely related to plants in the Ericales (an order of plants that includes heathers and azaleas).

Right: This alien-looking plant is *Mitrastemon matudae*, a poorly known parasite from Mexico. Once believed to be related to *Rafflesia*, it is now established to be more closely related to plants in the Ericales (an order of plants that includes heathers and azaleas).

HYDNORA
World's weirdest plant
Hydnora africana

....

This plant is surely one of the strangest of all flowering plants. Like the previous species, it lacks chlorophyll and steals its nutrients from the roots of other green plants. It attaches to the roots of succulent euphorbias in the dry semi-deserts of southern Africa, and lives entirely underground until it blooms, often unpredictably. The bizarre floral structures are about 20 cm (8 in) high, scarcely resemble a plant at all, and smell like faeces to attract pollinating dung beetles. Until recently its evolutionary origins were unclear. It is now established to be a distant relative of the Dutchman's pipes (*Aristolochia* spp., see pages 74–79).

Above: *Hydnora africana* in fruit.

Right: The bizarre flowering structure of *Hydnora africana*.

Following spread: The mouth-like blooms of *Hydnora africana* in the South African semi-desert, among their host plant, *Euphorbia mauritanica*.

MALTESE FUNGUS

False fungus

Cynomorium coccineum

....

This species looks more like a fungus than a plant, hence its common name. It is a rare and local species of dry habitats such as cliff-tops and desert sands in Mediterranean climate regions from the Canary Islands in the west to Afghanistan in the east (a related form of this plant also grows across Central Asia and Mongolia). The thick reddish spikes of tiny flowers are about 30 cm (12 in) high and are pollinated by flies, which are attracted to the flowers' unpleasant scent.

The Maltese fungus has a long history of use. Due to its distinctive form, it has long been used by herbalists as a remedy for sexual problems. It was highly prized by Maltese Knights; indeed it can still be found growing on Fungus Rock on the Maltese island of Gozo, now a designated nature reserve. DNA sequencing has revealed that the Maltese fungus is a distant relative of the saxifrages (family Saxifragaceae), a relationship hitherto unsuspected by botanists.

Right: *Cynomorium coccineum* grows on desert sands in the east of its range.

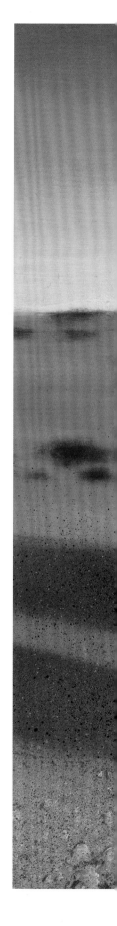

BALANOPHORA

Forest leech

Balanophora fungosa

....

The peculiar flowering structure of *Balanophora* looks much more like a fungus than a plant. There are about 20 species of *Balanophora*, which occur mainly in the moist forests of temperate and tropical Asia, extending to Africa, Madagascar and Australia. *Balanophora fungosa* is one of the most well-known species. It feeds from the roots of about 20 different tree species in the forests of Asia and Australasia. Like many parasitic plants, its leaves are reduced to scales, and completely lack chlorophyll. On the dingy forest floor, the plant is easily overlooked, and little is known about its biology. The evolutionary origins of *Balanophora* were a mystery until the advent of DNA sequencing technology. It is now known to belong to the order Santalales – a group of plants that also contains mistletoes. *Balanophora* is widely used in traditional medicine across the Asian continent to treat a variety of ailments.

Right: *Balanophora* (*Balanophora fungosa*) is a bizarre parasite that looks more like a fungus than a plant. It grows on the roots of rainforest trees in Asia and Australasia.

CYTINUS

Rootless, leafless vampire

Cytinus spp.

Cytinus, like tropical *Rafflesia* and *Rhizanthes*, has taken its pilfering lifestyle to extremes. This small plant is just a few centimetres across, spends the majority of its life embedded within its host plant's root system, and only emerges to flower. A handful of similar species are found in the Mediterranean Basin region and in South Africa where they parasitise the roots of various shrubs. As with other parasitic plants, the absence of true leaves confused early botanists about the identity of the closest relatives of *Cytinus*. DNA sequencing has now revealed that these plants do not belong to the *Rafflesia* family (Rafflesiaceae) as previously thought, but are in fact more closely related to the mallows (Malvaceae).

Above left: A large clump of *Cytinus hypocistis* growing on the roots of a *Halimium halimifolium* bush in southern Portugal.

Above right: *Cytinus* buds emerging from their host plant's root, a *Cistus monspeliensis*.

MISTLETOE

Tree-hugging vampire

Viscum spp.

Mistletoes are hemi-parasites. This means that unlike the previous species they do have green pigment (chlorophyll) and can photosynthesise; however they also extract water and nutrients from other plants. Mistletoes attach to the branches of their host trees as seedlings. The name mistletoe originally referred to the European species *Viscum album*, but is now used to describe various species from different families that share similar life histories. The seeds are dispersed by birds, which consume the fruits (technically known as drupes) and then excrete their droppings onto tree branches; the seeds are coated with an extremely sticky substance which hardens and fastens the seeds to the tree's branches, where they then germinate and attach.

Above left: Cape mistletoe (*Viscum capense*) growing in the low tree canopy of the Fynbos in the Western Cape, South Africa.

Above right: European mistletoe (*Viscum album*) in the tree canopy of Mount Olympus at 1,000 m (3,280 ft), North Greece, parasitising a black pine (*Pinus nigra*).

CISTANCHE

Vampire in the desert

Cistanche phelypaea

Cistanche is a genus of mainly desert-dwelling parasitic plants which attach to the roots of their host plants as seedlings. *Cistanche phelypaea* is a widespread species, occurring locally on dunes and desert sands from the Canary Islands to the eastern Mediterranean. It typically grows on the roots of shrubs in the Amaranthaceae family. In the spring, this plant sends up spectacular, leafless spikes of bright yellow flowers up to 1 m (3 ft) high, which illuminate the most barren parts of the desert. Close relatives of *C. phelypaea* have traditionally been used in Chinese herbal medicine. Because of their over-collection, the harvesting of their host plants for firewood, and the extreme difficulty of their cultivation, some species of *Cistanche* have suffered a dramatic reduction and are now endangered in the wild.

Above left: The leafless flowering stems of *Cistanche phelypaea* are completely devoid of chlorophyll.

Above right: An underground tuber of *Cistanche phelypaea* attached to its host root.

BROOMRAPE

Root vampire

Orobanche spp.

Broomrapes are related to *Cistanche*. There are hundreds of species of broomrape, which are most diverse across the Mediterranean Basin. Each species of broomrape has a preferred range of host plants and some grow exclusively on a single species. Broomrapes produce many thousands of minute dust-like seeds, which are wind-dispersed to increase the likelihood that some will fall within reach of a suitable host plant's roots. Some broomrapes have shifted from their natural host plant species to cultivated crops where they cause significant damage to agriculture. Others are extremely rare and threatened with extinction. New species of broomrape are still described every year. Broomrapes each have a preferred range of host plant. *Orobanche foetida* (above left) and *O. sanguinea* (top right) typically grow on just a few specific species in the pea family (Fabaceae).

Above left: *Orobanche foetida*, south Portugal.

Top right: *O. sanguinea*, Crete.

Bottom right: A spider-like developing broomrape attached to its host root.

DODDER

Vampire lasso

Cuscuta spp.

Dodder, also known as strangle-weed and witch's hair, not only steals from other plants, it lassoes and strangles them. When dodder seeds germinate they must find a suitable host plant within a few days or they will perish. Upon encountering a preferred host plant, they wrap their stems around it in a way similar to bindweed (*Convolvulus* spp.) which, indeed, is a relative of the dodder. They then penetrate the stem of their host, from which they extract water and nutrients. By the end of the growing season, these parasites can completely smother their hosts in great mounds several metres across. Some tropical species are so vigorous, they even reach the tree canopy. Found throughout the temperate and tropical world, like the broomrapes, some dodders are a severe constraint to crop cultivation.

Above: Masses of dodder stems (*Cuscuta approximata*) smothering their host, in this case, a thorny shrub in the Canary Islands.

DEVIL'S GUTS

Strangler

Cassytha spp.

'Devil's guts', also used to describe dodders, is the descriptive name for a family of over 20 parasitic plants native to Africa, Australasia, Asia and the Americas. Although they look very much like the dodders, they are in fact related to trees in the laurel family (Lauraceae). Independently, they have evolved a similar life history and appearance to the unrelated dodders – a process known as convergent evolution.

Above: The yellow thread-like stems of parasitic *Cassytha filiformis* smothering its host tree on the slopes of Table Mountain, South Africa.

KILLERS

....

Carnivorous plants have intrigued and inspired generations since Charles Darwin made the first detailed observations on them in the 19th century. These green predators have evolved in environments where nutrients are scarce, for example in waterlogged swamps or on rain-leached mountain slopes. They have evolved a bewildering array of lures and traps to attract, trap and digest animal prey to supplement their diet under these adverse conditions. There are nearly 600 species of them, and it is now known that they evolved independently several times. Their traps range from simple structures, such as tightly bound leaf rosettes, to intricate structures which employ unique strategies to capture and kill specific types of prey.

Left: *Nepenthes sibuyanensis* is a carnivorous pitcher plant native to Sibuyan Island in the Philippines where it grows on just two steep and remote mountainsides, and the ridge that connects them. It was first described only in 1996 and since then a plethora of new pitcher plant species have also been identified.

KING PITCHER PLANT

King of the pitcher plants

Nepenthes rajah

....

This plant is undoubtedly the most spectacular of all the carnivores: the king of the pitcher plants. First collected in 1858, it has fascinated and intrigued generations of botanists and enthusiasts ever since. Its gigantic pitchers can exceed 40 cm (16 in) in length, hold several litres of fluid, and were recently discovered to serve as 'tree shrew toilets' in much the same way as Low's pitcher plant described on page 40. Tree shrews that inadvertently slip into the pitchers and drown probably led to the commonly held view that this plant is capable of consuming rats. The plant grows only on Mount Kinabalu and neighbouring Mount Tambuyukon in North Borneo where it is threatened by collecting and is strictly protected.

Above: A newly opened king pitcher plant growing on Mount Kinabalu, North Borneo.

Right: The magnificent king pitcher plant in its natural habitat on Mount Kinabalu.

LOW'S PITCHER PLANT

Tree shrew toilet

Nepenthes lowii

....

This species grows on just a few mountainsides in North Borneo and is a very peculiar-looking pitcher plant. It was first described by the naturalist Hugh Low, who discovered it on his ascent up Mount Kinabalu and noted the unusual shape of the pitchers, which are vaguely reminiscent of toilet bowls. The pitchers on the plant's vines are up to 28 cm (11 in) long, are woody, and have a large opening. Relatively recently it was discovered that this species acquires nutrients in a way quite different to that of most insect-eating pitcher plants: the shape and size of the pitcher mouth matches the dimensions of tree shrews (*Tupaia montana*), which whilst feeding on the nectar on the inner surface of the spoon-like lid, excrete directly into the pitcher. The tree shrews' faeces provide the plant with essential nutrients. Therefore the peculiar toilet-like shape of the pitchers of this magnificent species is no coincidence!

Right: The peculiar, toilet-shaped pitchers of Low's pitcher plant (*Nepenthes lowii*).

Following spread: *Nepenthes lowii* in the remote, nutrient-poor cloud forests of Mount Trus Madi, North Borneo. Tree shrews sit astride the pitchers to feed on nectar on the inner surface of the lid. Whilst feeding, the tree shrews excrete into the pitcher vessel, which has glands to absorb the valuable nutrients the animals provide.

NEPENTHES EXTINCTA

Forever lost

Nepenthes extincta

....

New species of *Nepenthes* are described each year, but many are threatened by extensive habitat destruction. This painting of a carnivorous pitcher plant first described as *Nepenthes extincta* is based on a single herbarium specimen; this is probably all that remains of this plant. Some scientists believe it to be a natural hybrid of two other species of *Nepenthes*, but this is difficult to ascertain now that the plant has been wiped out. The plant formerly grew on Red Mountain in the Philippines, which is now the largest nickel-mining site in the world's second largest nickel-producing country. Today it is a sad emblem of the precarious continued existence of these beautiful plants in their natural habitats. Botanists are in a race against time to discover and conserve new species before they are lost forever.

Right: *Nepenthes extincta* once grew on Red Mountain in the Philippines – now a nickel-mining site. This was painted from a herbarium specimen, which is sadly all that remains of this plant.

MARSH PITCHER PLANT

Pitcher of death

Heliamphora spp.

....

Marsh pitcher plants are structurally among the most simple of the carnivorous pitcher plants. Their modified leaves are essentially foliar tubes, up to 50 cm (20 in) high, which attract insect prey with nectar. Wettable hairs on the inner surface of the pitchers are extremely slippery, and lead insects to lose their footing and aquaplane into a pool of fluid at the base. Unlike the more elaborate pitcher plants in Southeast Asia that produce digestive enzymes, the marsh pitcher plants contain bacteria, which produce enzymes to help break down their prey.

All marsh pitcher plants grow on isolated rocky plateaus called tepuis in South America. These are rainy, nutrient-leached and bleak mountains that arise abruptly from the rainforest floor and are home to a fascinating and unique flora and fauna.

Right: *Heliamphora elongata* growing on the rain-soaked rocky tepuis (Ilú–Tramen Massif) of Venezuela.

47

TRUMPET PITCHER

The intoxicator

Sarracenia flava

....

Trumpet pitchers are a genus of carnivores that grow predominantly in fens and swamps around the south-eastern plains of the United States. The pitchers grow up to about 90 cm (35 in) tall and their lurid colours and sweet nectar attracts insects. Insects that alight on the pitchers lose their footing on the slippery, waxy surface around the rim and tumble into a pool of digestive juices. The yellow trumpet pitcher (*S. flava*) has even been shown to produce a paralysing narcotic in its nectar, which may intoxicate its insect prey, leading them to lose their footing. After falling into the pitchers, the drowsy insects are barricaded by a series of downward-pointing hairs that prevent them from escaping. They eventually drown in a pool of digestive enzymes.

Above: Trumpet pitchers produce brightly coloured traps laced with sweet nectar to attract insect prey.

Right: Trumpet pitchers in their natural habitat.

VENUS FLYTRAP

A plant that can count

Dionaea muscipula

The Venus flytrap is among the most famous of the carnivorous plants. This carnivore grows in the subtropical wetlands of North and South Carolina, and typically traps insect prey, although small amphibians and reptiles are reputedly also sometimes caught. The terminal part of the leaf forms a barbed snare about 5 cm (2 in) across, which is triggered by minute hairs on the inner surface. When prey crawls on its surface, if more than one trigger hair is touched within a few seconds, or the same hair is touched twice, the trap is sprung. This mechanism is a safeguard against energy wastage from closing on, for example, wind-borne debris. The rapid closure of the trap is caused by an action potential, much like that of a nerve. Further movement triggers the secretion of digestive enzymes. Once the edges of the trap have compressed together tightly, a sealed cavity forms a veritable 'plant stomach'. Like other carnivorous plants, this 'foliar feeding' strategy has evolved as an adaptation to survival in nutrient-poor environments. It is believed that the Venus flytrap evolved from a sticky flypaper trap similar to that of the sundews, driven by an increase in prey size providing greater nutritional value over time.

Above: Touched once, the Venus flytrap's jaws remain open. Touched again in quick succession, and the trap is sprung.

Right: Typically the Venus flytrap catches insect prey, although small amphibians and reptiles are reputedly sometimes caught.

SUNDEW

Living flypaper

Drosera spp.

The sundews are a large group of carnivorous plants that occur on every continent except Antarctica. They typically inhabit acidic and nutrient-poor fens and bogs, and supplement their diet with small insects. Sundews range from a few centimetres across to over a metre in height. All species produce leaves with numerous glandular tentacles, which secrete a sweet mucilage to attract and ensnare insect prey. Insects that land on the leaves stimulate the sticky tentacles to bend inwards. Upon struggling to escape, the insects become exhausted and are asphyxiated by the tentacles' sticky mucilage. The insects are partially dissolved by digestive enzymes, and the nutrients subsequently released are then absorbed by the plant.

Above left: The dewy leaves of the sundew are irresistible to unsuspecting insect prey.

Above right: Insects find the dew-like secretions on the stalked glands of the sundew's leaves irresistible.

PARROT PITCHER PLANT

Lobster pot traps

Sarracenia psittacina

This bizarre plant is native to the wetlands of the Southeastern United States. Its traps have a very small entrance, concealed beneath the inflated hood. Insects that enter the chamber are unable to find the escape route – a mechanism known as a 'lobster pot trap', which is effective at capturing ground-dwelling insects. Within the trap, insects may be disoriented by light shining through translucent spots on the opposite side of the pitcher; as they try to escape through these false exits, they venture further into the trap. Eventually they crawl into the narrow tube at the base where their exit is impeded by downward-pointing hairs. Exhausted, they end up in a pool of digestive juices.

Above: Parrot pitcher plants can be completely submerged by water in their natural habitat, so it is possible that they may capture underwater prey – even tadpoles!

SECTION 3

FRAUDSTERS

Plants have evolved a range of elaborate mimicries to dupe, baffle and manipulate their pollinators. Darwin was fascinated by the various 'contrivances by which orchids are fertilised by insects'. The Orchidaceae is one of the largest of all plant families with an estimated 26,460 species. Many of these have evolved exclusive and often bizarre relationships with specific species of pollinating insect including bees, wasps, flies and moths. Most insect-pollinated flowers in the plant kingdom offer a reward, such as sugary nectar. Many orchids on the other hand deceive their pollinators with the false promise of a reward. Forms of deception range from flowers that advertise nectar but have none, through to the complex mimicry of a seductive female insect to dupe sex-crazed male bees. These unique strategies for deception in orchids, and the other plants described in this section, have all evolved to promote cross-pollination.

Left: *Psychotria elata*, native to Central and South America is sometimes called 'hooker's lips'. The lip-like structures are in fact a pair of bracts (the calyx) from which the true flowers later emerge. The peculiarly accurate resemblance to luscious human lips is not an elaborate form of mimicry like that of many orchids; the flowers actually attract pollinating hummingbirds.

ROTHSCHILD'S SLIPPER ORCHID

Charading as a brood of aphids

Paphiopedilum rothschildianum

This orchid is extremely rare. There are only a few places left on its native Mount Kinabalu in North Borneo where it still grows, and it has suffered a rapid decline due to over-collection. Its stems have fetched the highest prices of any flower, adding to its demand and, sadly, to its demise. Thankfully, with the advancement of propagation techniques, its outlook is now more positive. Rothschild's slipper orchid produces distinctive flowers with long lateral, outward-spreading sepals about 10–15 cm long (4–6 in). In common with many orchids, Rothschild's slipper orchid dupes its insect pollinators with the false promise of a reward. The sepals are covered in unusual clusters of small dark green spots that superficially resemble aphids. Parasitic wasps, which typically lay their eggs among broods of aphids, are attracted to the flowers as a potential breeding site. Female wasps lay their eggs on the flowers and subsequently fall or fly into the shiny pouch-like structure. Once inside, the wasps' exit is obstructed by in-turned lobes and they are forced to leave by an obscure escape route at the base of the flower. During this process, the wasps deposit pollen from previous Rothschild's slipper orchids they have visited onto the receptive stigma, and are in turn smeared with the plant's own pollen as they struggle to escape.

Right: Rothschild's slipper orchid is an exceptionally rare plant that grows only on the slopes around Mount Kinabalu in North Borneo.

TOAD CACTUS
Playing dead

Orbea variegata

....

The toad cactus is a mat-forming succulent that grows in the rocky deserts of South Africa (and is not related to the true cacti of the Americas). Insects such as bees, which are attracted to other brightly coloured and fragrant flowers, are often scarce in the desert. However, this plant attracts a different sort of pollinator altogether: carrion flies.

The flowers are about 7 cm (3 in) across, mottled, warty and smell strongly of carrion. The mimicry of dead flesh is so convincing that visiting flies frequently even lay their eggs on the surface of the flowers. During the course of egg-laying on the flowers, the insects pick up and deliver pollen, and bring about cross-pollination. After fertilisation, the plant sends up twinned, spike-like pods that split and shed wind-dispersed seeds far and wide across the desert.

Right: The unusual flower of the toad cactus.

STARFISH FLOWER

Desert starfish

Stapelia spp.

. . . .

The starfish flower is a close relative of the toad cactus. Like its relative, the starfish flower produces flowers that are attractive to pollinating flies. The mimicry of its flowers is so precise that they are even covered in a pelt of soft white hairs, resembling animal fur or mould. The flowers' unusual resemblance to a starfish – both their shape and their ground-hugging habit – is quite extraordinary, and they are a striking feature of the dry and barren mountain slopes of southern Africa.

Above: Flies are attracted to the smell of rotting flesh and search for a place to lay their eggs on a *Stapelia gariepensis* flower.

Right: Starfish flowers (*Stapelia gariepensis*) growing on dry mountains near the border between South Africa and Namibia.

61

FLYING DUCK ORCHID

Plant duck decoy

Caleana major

....

The flying duck orchid is found in southern and eastern Australia. Its remarkable little flower is about 2 cm (3/4 in) long and has a peculiar resemblance to a duck in flight. The flowers of this orchid are attractive to male sawflies. The sawflies mistake the flowers for female insects and attempt to mate with them, just like those of the bee orchids in the Mediterranean (described on page 66). However, the pollination mechanism of the flying duck orchid is more complex. The modified lip of this orchid comprises a hinged trigger mechanism. Vigorous male sawflies alight on the hinged lip and try to fly off with what they presume to be a female insect. During their struggle to take off with the decoy insect, they are repeatedly flung headfirst into a column where the reproductive parts of the flower are positioned. During this process, the pollen, packaged in bundles called pollinia, becomes glued to the insects' backs. The sawflies eventually fly off and the pollinia they carry undergo an automatic bending process. Thus, the pollen is primed into position for the next flower the insect visits to bring about cross-pollination.

Right: The flying duck orchid has a peculiar resemblance to a duck in flight.

MONKEY-FACE ORCHID

Monkeying around

Dracula simia

....

The monkey-face orchid (*Dracula simia*) is one of
a handful of orchids native to Central and South
America that, to the human eye, bear a striking
resemblance to a monkey's face. Research into a close
relative of this species shows that the rounded lower
lip of the flower mimics forest-dwelling mushrooms,
both in appearance and smell, to attract fungus gnats,
which pollinate the flowers.

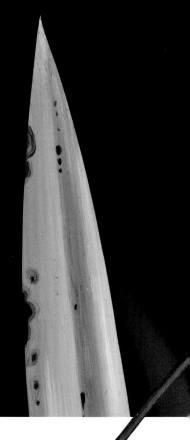

Right: The monkey face orchid
has a striking resemblance to a
monkey's face.

BEE ORCHIDS

Pollination by sexual deception

Ophrys spp.

Bee orchids (*Ophrys* spp.) are native to the Mediterranean Basin. These are the ultimate floral fraudsters: the flowers not only look just like bees, they even smell like the female of a particular species of bee. Amorous male bees which attempt to mate with the flowers inadvertently pick up and deposit pollen. How did this elaborate sexual deception strategy evolve? Each species of bee orchid targets a specific type of insect. This is very efficient because each dedicated pollinator achieves maximum pollen transfer, and valuable pollen is not wasted by being taken to other flowers. The orchids produce compounds virtually identical to the pheromones of female insects, and in the same relative proportions, to enhance the mimicry. Furthermore, this chemical mimicry evolved from leaf waxes – compounds which are common among plants. This duplicitous form of pollination is both effective and very efficient.

Above: There is a remarkable diversity of bee orchids in the Mediterranean, shown here by *Ophrys sphegodes* subsp. *mammosa* (top left), *O. cretica* (top right) and *O. scolopax* subsp. *heldreichii* (bottom right). They each target specific species of male bees, which bring about cross-fertilisation whilst attempting to mate with the flowers.

TONGUE ORCHID

Pollination by sleepy bees

Serapias spp.

The tongue orchid produces flowers with a tongue-like lip, which is enclosed at the base to form a tunnel. Tongue orchids are closely related to bee orchids and also occur in the Mediterranean Basin. Pollination occurs during the afternoon when solitary bees hover from flower to flower. The bees eventually rest inside the tunnel of one of the flowers and remain there overnight, or during wet weather. The bees that visit the tongue orchids typically seek refuge in holes in the ground, so the tongue orchids may have evolved to mimic these protective shelters. It has also been suggested that the male bees may search for female insects within the floral tubes of the flowers. Thus the flowers may in fact be mimicking a nest entrance, or a flower with a female insect inside, to entice male bees. This subtle pollination mechanism appears to be yet another complex deception strategy to promote cross-pollination in orchids.

Above left: A stand of tongue orchids (*Serapias orientalis*) in Cyprus.

Above right: A small bee, recently emerged from a tongue orchid flower, laden with packets of pollen (pollinia).

SECTION 4

JAILERS

Plants have evolved prison chambers to incarcerate pollinating insects. Aroids belong to a large family (Araceae) of over 3,300 species, which is especially diverse in the New World Tropics. The distinguishing feature of the aroids is a spike-like spadix that is often partially enclosed by a bract-like spathe. The true flowers are typically minute and develop on the spadix, with the female flowers at the base and the male flowers immediately above. The female flowers ripen before the male flowers – a time switch that prevents self-fertilisation. Interestingly, many plants in the family are heat-producing (thermogenic), their blooms becoming much warmer than the surrounding air. The combination of heat and a powerful smell, often of animal dung or of rotting flesh, is very attractive to insects such as beetles or flies, which may mistake the plant for carrion. In many species, the spathe encloses the minute flowers in a floral chamber into which insects seeking a place to lay their eggs become trapped. This mechanism ensures insects pick up and deliver pollen. This form of 'trap pollination' has also evolved independently in some other plants, for example in *Hydnora* (page 22) and the Dutchman's pipes (family Aristolochiaceae). A small selection of these trap pollinators is described in this section.

Left: *Arisaema nepenthoides*, an aroid native to the eastern Himalayas.

MEDITERRANEAN ARUM

Dung fly jail

Arum pictum

After the first heavy autumn rains on the central Mediterranean islands, the Mediterranean arum sends up its curious floral structures from the pine forest floor. The blooms are about 25 cm (10 in) high, brownish-purple, and smell strongly of horse dung, which is attractive to the little midges that typically teem around animal droppings. To perfect its mimicry of dung, the spike-like spadix is even warm to the touch, just like fresh animal faeces, which helps to disperse the odour. The little insects are imprisoned in the basal floral chamber overnight by a barricade of downward-pointing spines, and become showered with pollen. The following day, the spines wither to release the insects, some of which will visit another Mediterranean arum and bring about cross-pollination.

Above: A cross-section of a Mediterranean arum revealing the floral chamber in which insects are trapped by a barricade of downward-pointing spines and showered in pollen shed by the male flowers (the yellow central band). The greenish female flowers are at the base.

Right: The curious bloom of the Mediterranean arum.

DEAD HORSE ARUM

24 hour pollination sentence

Helicodiceros muscivorus

....

The dead horse arum grows high on cliff-tops
overlooking the Mediterranean among sea gull colonies.
Here, the odour of regurgitated fish, gull droppings
and dead chicks leads to an abundance of bluebottle
flies. The dead horse arum's 40 cm-long (16 in) blooms
drape over the rocks and resemble animal corpses in
both appearance and smell to capitalise on this bounty
of flies. The flies swarm around the plant looking for
a place to lay their eggs. Upon crawling into a hollow
chamber, they become trapped by stiff backward-
pointing slippery spines and are showered in pollen.
The next day, the spines wither and release the flies,
some of which will visit another dead horse arum and
bring about cross-pollination.

Right: A dead horse arum,
awaiting pollinating flies, by the
Mediterranean Sea.

DUTCHMAN'S PIPE

Suspended sentence

Aristolochia spp.

Dutchman's pipes, also known as birthworts, are a diverse group of plants that occur on most continents. The flowers are typically held suspended on vines and are attractive to small flying insects. Upon alighting on the lip of the flower, the insects crawl into a narrow floral tube, which is clothed in dense downward-pointing hairs. These hairs are easy to crawl down, but very difficult to crawl up in the opposite direction, leading the insects to become imprisoned in the suspended floral chamber. The hairs later wither, unblocking the flower's escape passage, to release the pollen-laden insects, which are then primed to cross-pollinate another Dutchman's pipe flower.

Above: *Aristolochia baetica*, scrambles among rocks and cliffs in the south-west Mediterranean.

Right: The pelican flower (*Aristolochia grandiflora*) is a species from Central America that has one of the largest flowers of the New World, up to 60 cm (24 in) long.

Page 76: *Aristolochia didyma*, an unusual species of Dutchman's pipe from South America.

Page 77: Dutchman's pipes (*Aristolochia* spp.) have evolved an extraordinary diversity of blooms. These flowers belong to the species *A. tricaudata*, a vine native to Mexico, and have the rather gruesome appearance of deep sea monsters.

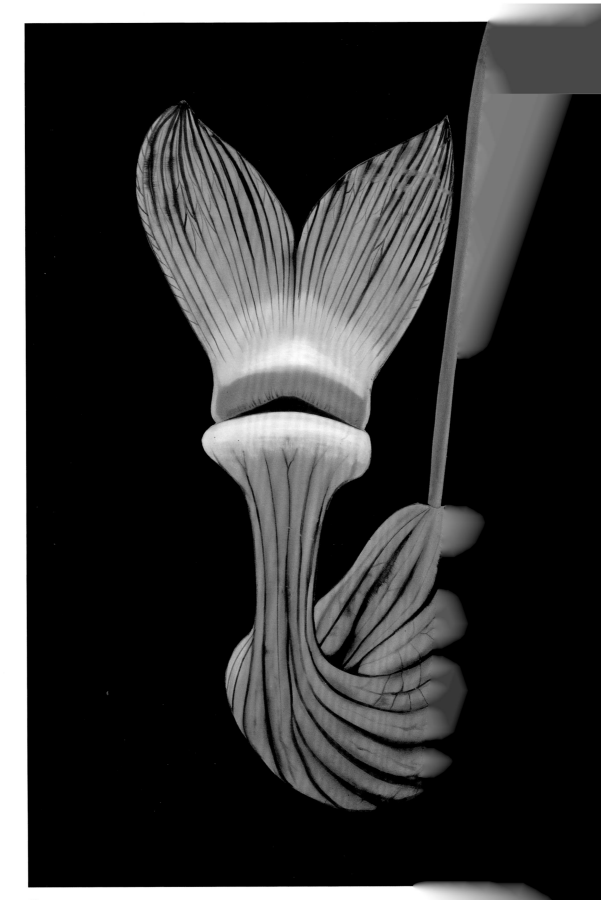

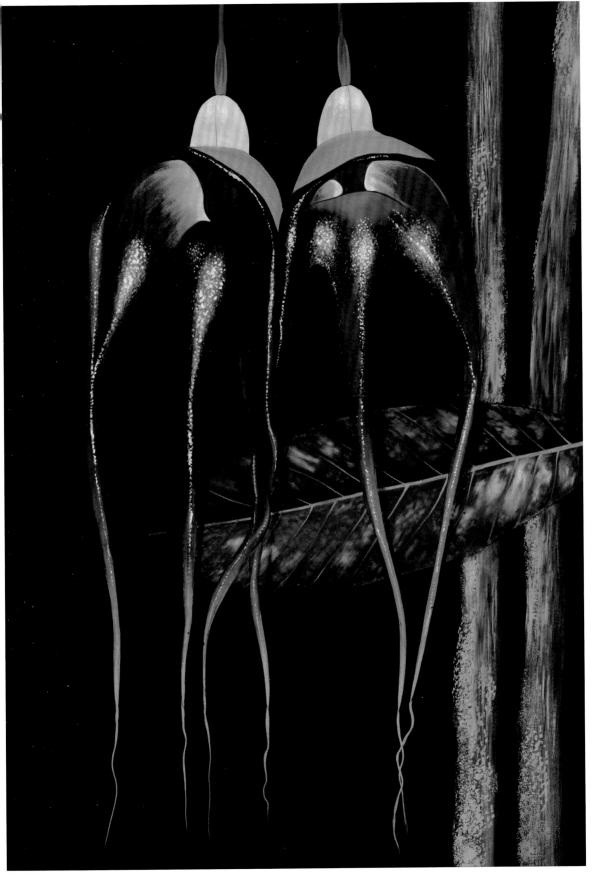

ARISTOLOCHIA ARBOREA

Masquerading mushrooms

Aristolochia arborea

....

Aristolochia arborea is a spectacular species of Dutchmen's pipe that grows in the rainforests of Central America. The plant has a shrubby, tree-like habit. Its peculiar blooms sprout directly from the very base of the trunk and resemble clusters of mushrooms which grow on the rainforest floor. In fact, each *Aristolochia arborea* bloom has a highly specialised floral outgrowth in its centre that resembles a little purplish-brown toadstool. It mimics with incredible accuracy a specific form of toadstool that grows in the surrounding forest, and even emits a mushroom-like smell. Just like the monkey-face orchid (see pages 64–65), it is attractive to fungus gnats, which typically lay their eggs on toadstools. The gnats are so convinced that the blooms are toadstools, they lay their eggs on the plant. Whilst they lay their eggs, they pick up and deliver pollen; meanwhile their offspring, which are unable to feed on the bogus toadstool, will perish.

Right: The blooms of *Aristolochia arborea* closely resemble a particular form of toadstool to attract pollinating fungus gnats, which are so convinced by the plant's mimicry, they even lay their eggs on it!

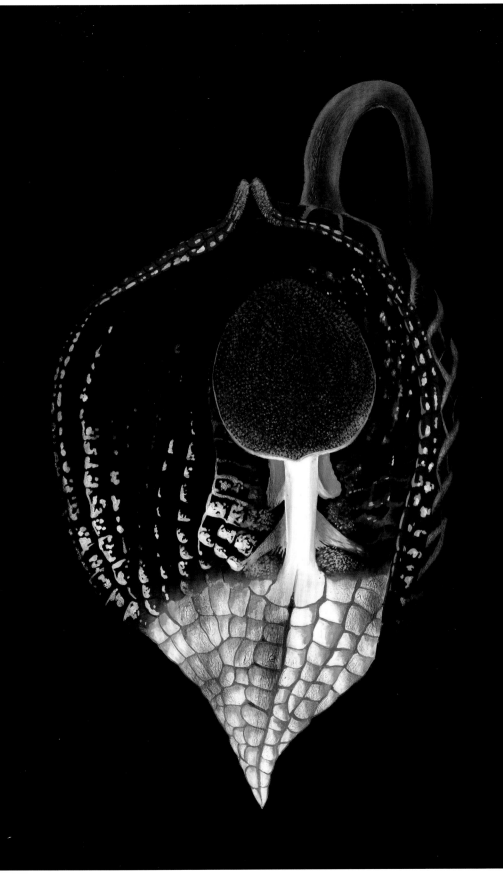

TITAN ARUM

King of the jungle

Amorphophallus titanum

This giant of a plant, which only grows in the forests of Sumatra, has become world-famous for its monstrous size. Because its bloom comprises a collection of tiny flowers (an inflorescence) it is not strictly speaking the world's single largest flower (see *Rafflesia*, page 14). Nevertheless, the titan arum can reach a height of 3 m (10 ft) and is considered to have the largest unbranched inflorescence of all plants. It is a distant relative of the Mediterranean arum and dead horse arum described previously (pages 70–73), and like these plants it mimics a corpse to attract insects that feed on carrion. When the purple spathe opens to reveal the towering yellow spadix, the female flowers are receptive to pollen from other titan arums. Insects that may have visited a bloom elsewhere crawl around the base of the inflorescence and spread pollen all over the female flowers, bringing about cross-fertilisation. The male flowers later mature and shed pollen over visiting insects. Chemical analysis shows that the compounds released from the spadix of the titan arum resemble those that emanate from sweaty socks and rotting fish! After flowering, it can take 7 to 10 years before the plant blooms again.

Right: The gigantic bloom of the titan arum.

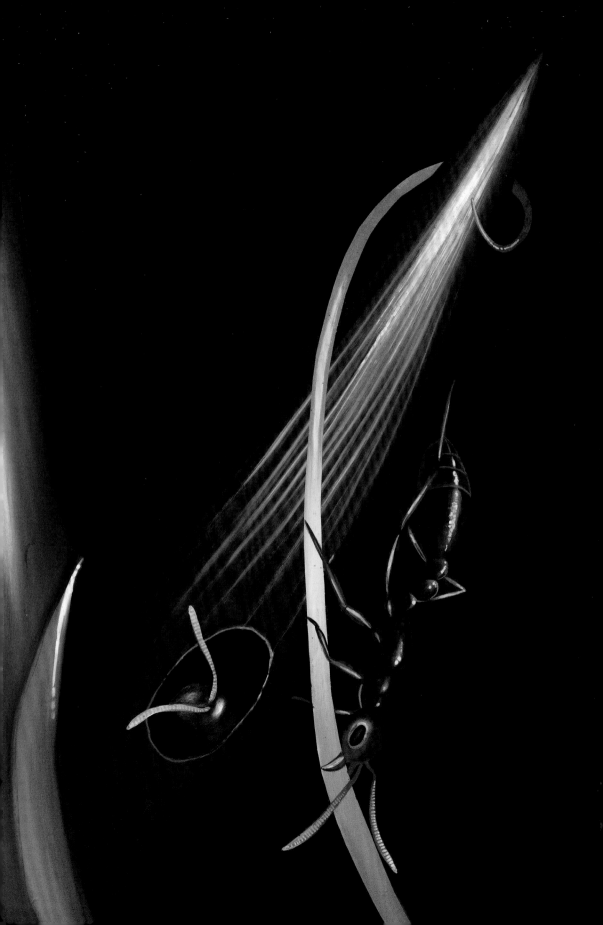

SECTION 5

ACCOMPLICES

Animals are accomplice to a diversity of intricate and surprising schemes which plants have evolved to bring about their pollination services, their dispersal, or even for self-defence. Tropical bucket orchids, for example, supply male bees with a perfume they use to attract females, meanwhile sending them on a gruelling pollination assault course, leaving them bewildered and bedraggled. Other plants employ marching ant armies to patrol their stems and defend them from attack by predators, in exchange for a nesting site. More surprising still, a handful of the voracious killer pitcher plants have evolved mutualistic relationships with animals. For example roosting bats nourish one species of pitcher plant with their droppings, in return for a safe place to rear their young. Many of these mutualistic relationships that have evolved between plant and animal are so intertwined that neither could exist without the other.

Left: Ant plants produce modified, hollow nesting structures called domatia, which are adapted for habitation by ants. In this African acacia, the domatia are derived from hollow thorns. In exchange for accommodation inside the thorns, the plant's partner ants patrol the tree's branches and attack would-be herbivores and competing plants, biting through their tendrils.

BUCKET ORCHID

Pollination obstacle course

Coryanthes macrantha

The aroids and the Dutchman's pipes (pages 70–79) trap insects and separate their male and female reproductive parts over time, that is, the female flowers usually ripen before the male. The South American bucket orchid also traps its pollinators, but this orchid separates its male and female reproductive parts spatially. The flowers, which are about 12 cm (5 in) across, are visited by particular species of male euglossine bee (Family Apidae). The male bees harvest fragrant compounds produced by the flowers, which they use to attract female bees to mate. As the male bees teem around the flowers, some inevitably fall into a fluid-filled 'bucket' – the modified lip of the flower. The inner surface of the bucket is waxy and prevents the insects from gaining a grip. However, a ridge on the surface provides a foothold and directs the bees towards a tight passageway of just the right dimensions for the insects to squeeze through. As the bees push through this passage, they first pass the receptive female part (the stigma). If an insect happens to be carrying packets of pollen (pollinia), picked up from another bucket orchid, these are deposited, cross-fertilising the flower. As the insects struggle further through the passageway, pollinia from the male part of the flower are then glued to their thorax. The pollinators eventually escape and may go on to cross-pollinate other bucket orchids.

Right: The peculiar flower of the bucket orchid.

BIRD OF PARADISE

Pollination springboard

Strelitzia reginae

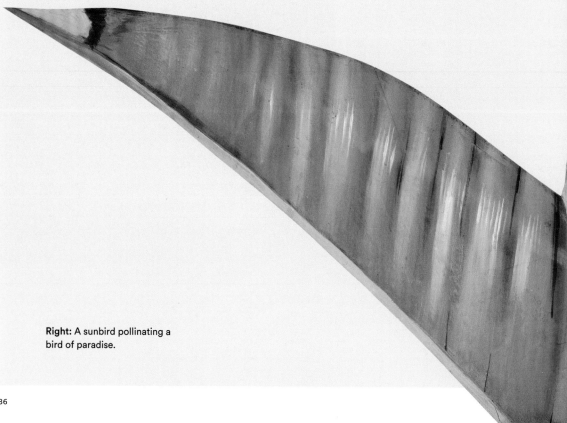

The bird of paradise plant is a perennial that grows to about 2 m (6.5 ft) high in the Fynbos vegetation of the Cape Provinces of South Africa. The beak-like spathe and prominent orange sepals resemble a bird's head. The flower's structure is a complex mechanism, which ensures that when a sunbird perches on the blue arrowhead-like petals, they open up and thrust their anthers outwards to dust the bird's chest with pollen. In fact, most birds cling to the side or tip of the flower to avoid exactly this, to prevent soiling their plumage.

Right: A sunbird pollinating a bird of paradise.

ANT PLANTS

Ant empires

Myrmecodia spp.

....

Ant plants produce specialised hollow nesting structures called domatia, which are specifically adapted for habitation by ants. Among the most elaborate structures are those produced by Southeast Asian *Myrmecodia* species, such as *M. lamii* in West Papua. These plants produce swollen, spiny domatia at the base of their stems. The domatia, which are about 40 cm (16 in) across, contain a complex labyrinth of hollow cavities that are connected to the exterior by entrance holes. Smooth-walled chambers are used by the ants for nesting, while rough, warty cavities serve as waste chambers in which ants deposit nutrient-rich faeces. The ant plants are epiphytic, meaning they grow perched high on the branches of trees where the ants provide a valuable source of nutrients, which are otherwise scarce.

In addition to domatia, some species of ant plant also offer their partner ants food rewards such as nectar and specialised nutrient-rich packets. Some species of ant even tend 'gardens' of sap-feeding insects (similar to aphids) on the plants, which yield honeydew that they feed upon. Another example of 'agriculture by ants' is the encouragement of fungal growth within the domatia. These fungal-farming ants provide nutrients for the fungi and use them as a food source for their larvae in return. It has been suggested that these fungi may also enhance the transfer of nutrients to the plant.

Right: *Myrmecodia lamii*, an ant plant that often grows perched on tree branches as an epiphyte in the cloud forests of West Papua.

FANGED PITCHER PLANT

Ant plant carnivore double act

Nepenthes bicalcarata

....

The fanged pitcher plant attracts, traps and digests invertebrate prey, just like the other pitcher plants described previously. But unlike other pitcher plants, this Bornean species has evolved a close relationship with a specific species of ant. The ants rear their young in the plant's coiled, swollen, hollow tendrils. The ants consume the copious nectar produced around the fangs that project beneath the lid (from which the plant derives its name). In return for food and accommodation, the ants viciously attack weevils that attempt to feed on the developing shoots of the pitcher plant. Interestingly, the ants, unlike insect prey, can dive in and out of the pitcher fluid with impunity. It has even been suggested that the ants may drag out larger prey items that could potentially cause the pitchers to rot if left inside. Mutualistic relationships with ants have evolved elsewhere in the plant kingdom (see ant plant, page 88), but among the carnivorous plants, this form of symbiosis appears to be unique to the fanged pitcher plant.

Right: The fanged pitcher plant is home to a particular species of ant, which lives in the plant's swollen tendrils.

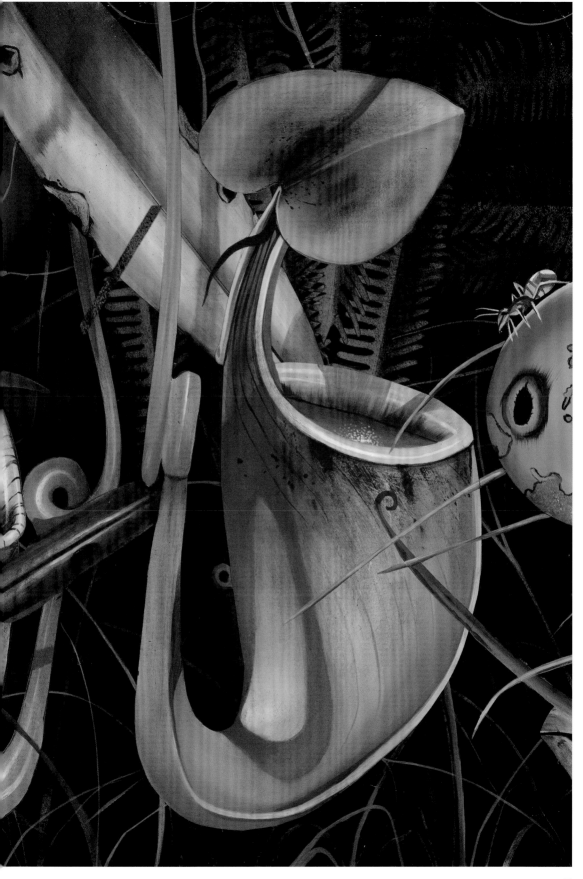

BAT PITCHER PLANT

Bat bed & breakfast

Nepenthes hemsleyana

Like Low's pitcher plant and the king pitcher plant, already described (see pages 38–43), the pitchers of this Bornean species capture animal faeces, but in a radically different way. The plant's long, slender pitchers lack the strong fragrance and copious nectar produced by most pitcher plants, and so this species traps fewer insects. Remarkably though, the pitchers of this plant provide a daytime roosting site for woolly bats (*Kerivoula hardwickii*). The pitchers have a prominent ridge onto which the bats can cling, and an enlarged opening that effectively reflects the ultrasound calls of the bats, enabling them to locate the plant among dense vegetation. Indeed, the pitchers have a striking resemblance to the bat-pollinated flowers of other plants, which are usually tubular, white or pale green and suspended on long stalks. This incredible relationship benefits both the bat, which uses the pitchers as a safe roosting site, and the plant, which is nourished by the bat droppings in the nutrient-poor heath forests where it grows.

Right: Woolly bats make their home in bat pitcher plants.

NEW ZEALAND FLAX

Reptile pollination

Phormium tenax

....

In New Zealand, geckos feed on the sugary nectar of a variety of flowering plants. Duvaucel's geckos (*Hoplodactylus duvaucelii*) typically congregate on trees infested with scale insects to exploit the honeydew they secrete. However, when flax plants come into flower, the geckos migrate long distances to visit these plants, prise open their petals and extract the nectar inside, using their long tongues. As they feed from the flowers, their chins and throats, which have modified scales, become dusted with pollen, which they may then transfer to other flowers, bringing about cross-pollination.

Right: A Duvaucel's gecko feeding from a New Zealand flax plant.

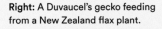

SURVIVORS

Plants have evolved to jostle with the elements in some of the most hostile places on our planet. For example even in the driest deserts, plants have evolved a suite of survival tactics far superior to those of any animal. Peculiar desert-dwelling *Welwitschia* sits out millennia of long droughts, only ever producing two leaves. Many desert plants have succulent leaves and stems which store water to survive long, hot, dry spells. Others, such as the Australian Sturt's desert pea, employ 'boom and bust' tactics, flourishing after rare rainy spells, and rapidly flowering and setting seed. It is seeds that make plants the ultimate masters of survival. Seeds can lie dormant for long periods of time, only geminating when conditions are right, in some cases for many thousands of years.

Left: *Welwitschia*, the ultimate desert survivor among plants.

HALF-MENS

Half-plant, half-man

Pachypodium namaquanum

....

Half-mens are iconic trees of the deserts of the Orange River region of South Africa's arid Northern Cape and southern Namibia. This plant owes its unusual name to the Khoikhoi myth claiming that the succulent is half-plant, half-man. The giant, pillar-like stems can reach a height of over 4 m (13 ft). The stems are covered with spines and crowned at the top with rosettes of leathery leaves, which are shed in the hot, dry summer months. The small quantity of leaves, which are easily shed, reduce the surface area from which water can be lost to evaporation. Furthermore, the swollen trunk has a significant water-storage capacity, enabling the succulent to survive even the longest desert droughts. Indeed, the plant is so adept for surviving these adverse conditions, it is believed to be capable of reaching an age of several centuries.

Right: The half-mens succulent, named after local legend; when viewed in silhouette, the plant is said to look like a person.

STURT'S DESERT PEA

Desert survival champion

Swainsona formosa

....

Sturt's desert pea (*Swainsona formosa*) is one of Australia's most iconic plants and a symbol of the country's desert-dwelling flora. It is native to the arid central and north-western regions where rainfall can be exceptionally scarce. Like many desert-dwelling survivors, this species has seeds that can remain viable in the barren desert sands for years until conditions become favourable for germination. After rainfall, the seedlings rapidly establish long roots so that the plant can flower and set seed in advance of the return of unfavourable hot, dry spells.

Right: The vibrant flowers of a Sturt's desert pea, which appear after rainfall in the desert.

WELWITSCHIA

Millennial two-leaf wonder

Welwitschia mirabilis

....

Welwitschia is a monotypic genus, meaning there is just one species – *W. mirabilis*. This unique plant grows in barren deserts along the Atlantic sea-belt of Namibia and Angola. Unlike the other species described in this book, which are all flowering plants, *Welwitschia* is a gymnosperm – a relative of the pines and cycads. This species is one of the ultimate desert survivors. Some individual *Welwitschia* plants are believed to have lived for over 1,000 years, in conditions of extreme drought. Perhaps the most unusual and unique features of *Welwitschia* is that it produces just two true leaves. These continue to grow throughout the long life of the plant, to a length of 2–4 m (6.5–13 ft), becoming split and frayed over time to form numerous strap-like segments.

Right: *Welwitschia*, the ultimate desert survivor among plants.

CANDELABRA LILY

A firework and a tumbleweed

Brunsvigia orientalis

Candelabra lilies lie dormant during the hot, dry summer months in the South African Cape, in the form of large, water-storing bulbs underground. After blooming in the autumn, the dry flowerheads detach from the bulbs and roll with the wind in the form of 'tumble weeds', scattering their seeds as they go. The seeds swiftly germinate during the damp winter months, before retreating from the return of the fierce summer sun below ground in the form of a bulb.

Above left: Candelabra lilies flower in the autumn, producing magnificent heads like fireworks in the Western Cape.

Above right: Candelabra lilies produce leaves in the winter and remain dormant during the hot summer months in South Africa's Western Cape.

COCO DE MER

Island survivor

Lodoicea maldivica

This survivor is a tropical island 'hanger on'. The coco de mer palm (*Lodoicea maldivica*) produces the largest seed in the plant kingdom. The hard, smooth, brown seeds are enormous, weighing up to an incredible 17 kg (37 lb) or more. It is native to the Seychelles, where it has become an icon for conservation, and is now a major tourist attraction. Prior to human colonisation the tree was dominant on the islands of Praslin and Curieuse. Sadly, human activity on the islands has drastically reduced the populations of this tree in the last two centuries, and there remain only three populations in the wild. The trees take a long time to re-establish; some mature trees are estimated to be up to 200 years old. Unlike some of the plants in this book, the coco de mer palm, unsurprisingly, has a very limited seed dispersal.

Above: The largest seed in the world is produced by the coco de mer palm, and weighs an incredible 17 kg (37 lb) or more.

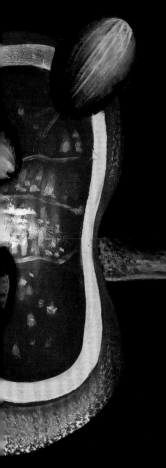

SECTION 7

HITCH-HIKERS

Plants have evolved an astonishing array of fruits and seeds to facilitate their dispersal. Some have parachutes which they employ for wind dispersal, or float on water, to travel vast distances across the world's oceans. Others are more unusual: the peculiar Southeast Asian durian fruit smells absolutely repulsive to most human noses. But by broadcasting this overpowering and unique stench, the durian attracts animals from far and wide to disperse its seeds. Fruits and seeds come in a bewildering variety of colours, shapes and sizes. Some are bright red to entice birds; others are a vivid shade of blue to attract the attention of lemurs. One fruit, to our eyes, looks remarkably like a coiled-up snake, hanging from a vine.

Left: The blue-black seeds of the peanut tree (*Sterculia quadrifida*) appear to float upon the fruit's contrasting vivid red flesh.

DURIAN

Rainforest hitch-hiker

Durio zibethinus

The durian has been described as the 'king of fruits'. It is highly prized in its native Southeast Asia, and renowned for its large size – up to about 30 cm (12 in) across – strong odour and unique flavour. Many people find the aroma, which has been compared with rotting onions and sewage, as overpowering and highly distasteful; so much so, it is frequently banned from public transport and hotels. The fruit's flesh, however, is prized for its sweetness. There are many cultivars and the fruits of some can fetch high prices.

The durian tree reaches a height of approximately 45 m (148 ft) and produces flowers that sprout directly from the trunk (a process known as cauliflory) and are bat-pollinated. The durian is a particularly effective rainforest hitch-hiker. Its strong smell is attractive far and wide to an array of forest-dwelling mammals including orangutans, hornbills and macaques, which swallow the seeds whilst eating the fruit's flesh. These animals are highly effective dispersal agents for the durian, covering considerable distances across the rainforest.

Right: The creamy flesh of the durian fruit encases large seeds which are often swallowed whole by a variety of rainforest animals.

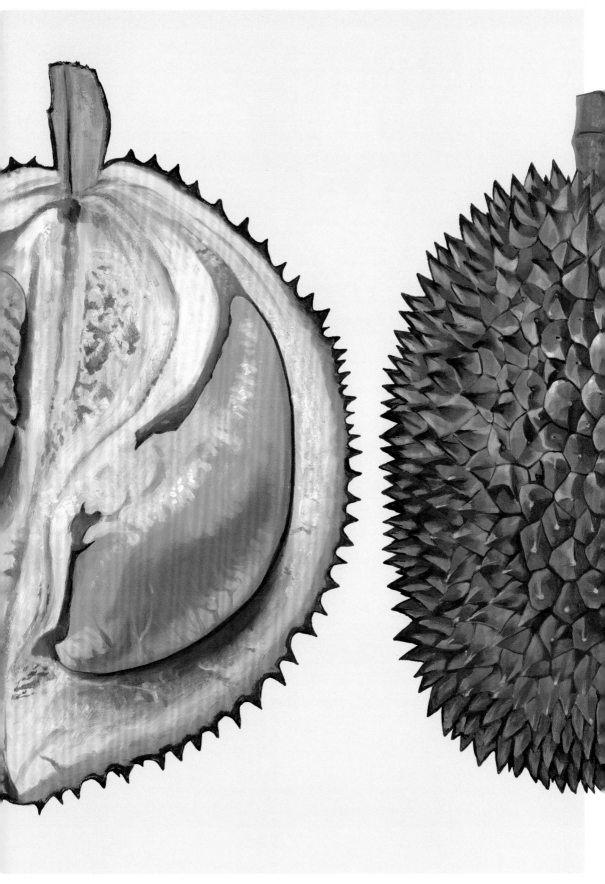

SNAKE GOURD

Snake or fake?

Trichosanthes cucumerina

....

Fruits and seeds – the dispersal agents of plants – come in all shapes and sizes. The snake gourd (*Trichosanthes cucumerina*) is a member of the pumpkin and melon family (Cucurbitaceae). However this particular tropical member produces eccentric, snake-like fruits. The fruit is remarkably long, often spanning over a metre, and turns reddish when mature. Some local growers tie rocks to the end of the developing fruit to straighten out the peculiarly serpentine coils the fruit produces if left to mature naturally. Owing to its unusual appearance and culinary use, this hitch-hiker has been spread by people and is now cultivated as a vegetable, and for traditional medicine, throughout much of the tropics, including southern Asia, tropical Africa and Madagascar.

Right: The curious, serpentine fruit of the snake gourd.

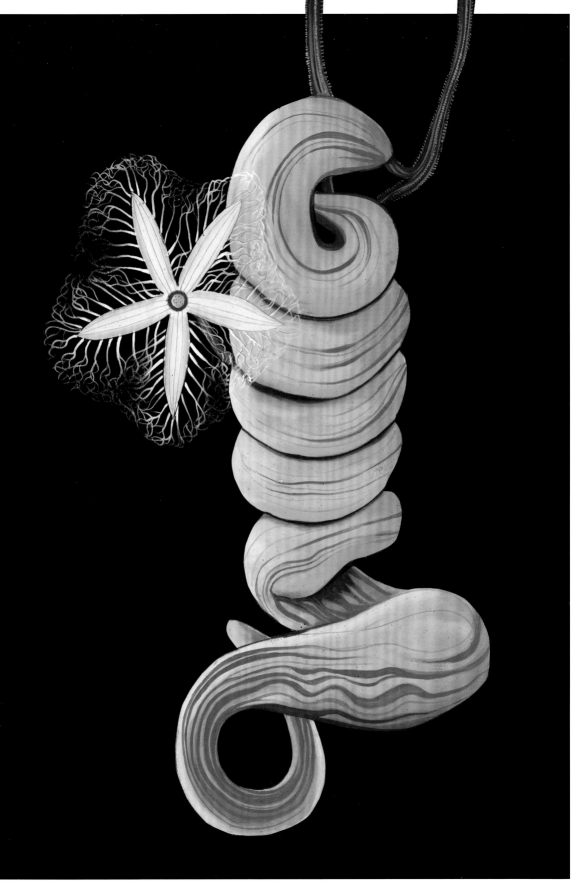

TRAVELLER'S PALM

Black and blue seeds

Ravenala madagascariensis

....

A close relative of the bird of paradise plant, this Madagascan endemic is not actually a true palm. The vibrant seeds are produced in capsules about 9 cm (3.5 in) long. The related South African bird of paradise produces seeds with red appendages, while the seed appendages of the Madagascan traveller's palm are a remarkable shade of blue. Red seeds and fruits are relatively common in nature, particularly among bird-dispersed species, but blue is exceptionally rare. The reason the traveller's palm has evolved blue seeds is linked to its biogeography. There are few fruit- and seed-eating birds on Madagascar, and instead, this species has evolved a mutualistic relationship with lemurs. Lemurs' vision can only differentiate shades of green and blue, unlike the vision of birds. Lemurs are therefore attracted to the bright blue, edible appendage of the traveller's palm's seeds and disperse them, to the advantage of the plant.

Right: The black seeds of the traveller's palm have a contrasting edible blue structure which is highly attractive to lemurs, which disperse the plant's seeds.

INDEX OF PLANTS

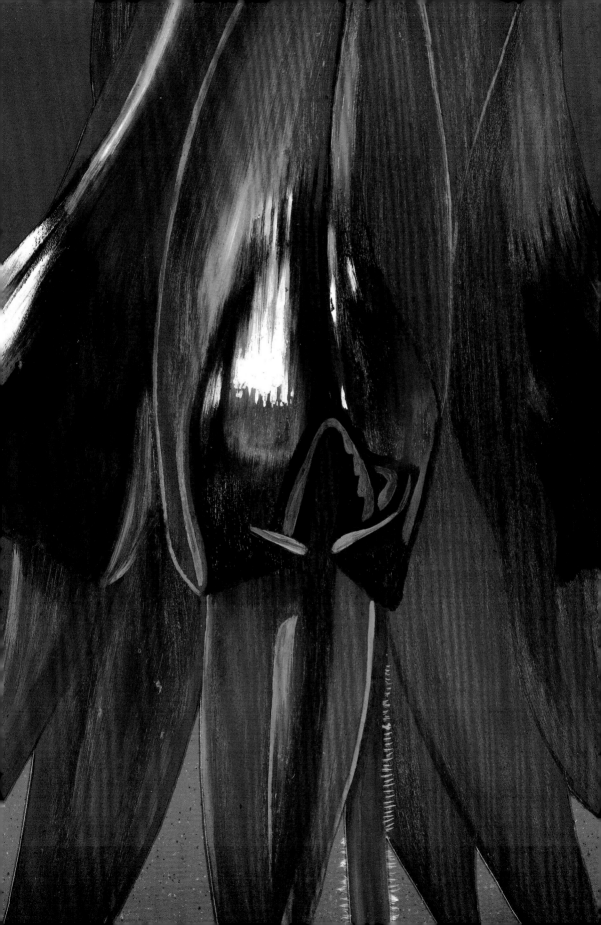

INDEX OF PLACES